VOLUME 57 JANUARY 2001 PAGES 1-28

QUINNIPIAC FISHES & FISHERIES

History and Modern Perspectives on the Fishes and Fisheries in the Quinnipiac Watershed

BY
JON A. MOORE

TRANSACTIONS
THE CONNECTICUT ACADEMY OF ARTS AND SCIENCES

P.O. Box 208211
New Haven, Connecticut 06520-8211

CAAS@YALE.EDU

INTRODUCTION

Connecticut is dominated by three major river drainages (the Connecticut, Housatonic, and Thames Rivers) that extend beyond the state's boundaries. Much attention and research has in the past focused on these river systems (McDonald 1887, Smith 1946, Hard 1947, Moss 1960, Hames 1975, Tolderlund 1975, Merriman and Thorpe 1976). Despite the fact that the Quinnipiac River watershed represents the fourth largest river system in Connecticut at 459.8 km^2 (Hagstrom et al. 1991) and passes through one of the most densely populated portions of the state, most investigators have largely overlooked it. However, many of the trends that will be discussed in reference to the Quinnipiac (i.e., early abundance of species, development of fisheries, dam building, extirpation of anadromous fish runs, pollution, and introductions of exotic species) are typical of what was occurring contemporaneously in other watersheds throughout the state.

The Quinnipiac River runs for 38 miles from its headwaters in Farmington to its mouth in New Haven Harbor (Fig. 1). There are about 20 tributaries; the largest of these include the Eight Mile River, Ten Mile River, Harbor Brook, Wharton Brook, and Muddy River. The watershed itself covers 170 square miles spread over 14 communities, particularly the towns of Southington, Cheshire, Meriden, Wallingford, North Haven, Hamden, and New Haven. The limits of the saltwater influence extend along the main channel of the Quinnipiac from the harbor up to the mouth of Waterman Brook in North Haven (O'Hare 1981), and this is taken to constitute the estuary.

Past Fishery Surveys in the Watershed

The Reverend James Linsley (1844), a Yale trained theologian turned natural historian, published the first scientific documentation of fishes found in Connecticut, a list of all marine and freshwater fish species found within the state. Linsley notes the occurrences of species in the Connecticut, Housatonic and Thames Rivers, but unfortunately makes no special reference to the Quinnipiac River. This omission is puzzling given that the lower reaches of the river are within a few miles of Yale. Rev. Linsley might have actually sampled fishes from the Quinnipiac, but considered it a minor watershed of no note.

The Quinnipiac River, however, was a part of the original New Haven Colony and as such represented a natural resource available to and used by the colonists. Early records from the colony do provide a little insight into the fishes and fisheries found in the Quinnipiac watershed. [Note: in early colonial and state records the Quinnipiac is called the "East River," and also occasionally the "New Haven River." The name "Quinnipiac River" apparently came into use in the early to mid-19th century.]

Despite the close proximity of Yale University, which has a long tradition in natural history since the beginning of the 19th century (Narendra 1988, Moore and Boardman 1991), it is only very recently that more extensive scientific collections and studies were made of the fishes in the Quinnipiac watershed. Professor Addison E. Verrill did catalogue a few common species of fishes, such as common shiner and sea lamprey, into the zoology collections of the Peabody Museum of Natural History at Yale University (YPM). According to museum catalogue records, these fishes were all collected from the Quinnipiac River in 1877. But no comprehensive collecting was conducted until Whitworth et al. (1968) reported on the fishes found at 14 sites in the Quinnipiac watershed during the 1960s. Information was compiled from personal collections, museum collections, state fisheries efforts, and other sources. The work provided the first summary of information on fishes found in the Quinnipiac watershed.

Since 1981, the Connecticut Department of Environmental Protection (DEP) Fisheries Division has been conducting surveys of lakes (since 1981; Jacobs and O'Donnell 1995) and since 1988 on streams and rivers (Hagstrom et al. 1991, 1994, 1995). These surveys have added significantly to our knowledge of the fishes in the Quinnipiac. Other recent activities in the Quinnipiac watershed include researchers from Yale University (G. Berghorn unpubl. data) and other organizations (Dunbar 1984; Jokinen 1994; Normandeau 1979 & 1985; Pereira et al. 1994; U.S. Fish & Wildlife Service unpubl. data).

Our knowledge of the fishes from the Quinnipiac watershed, therefore, divides into three time frames: 17th, 18th, and 19th century historical records; surveys or collections from the 1930s to 1970; and surveys or collections made after 1980. Unfortunately, information prior to the 1940s is very scanty, so we will draw conclusions only from the most recent changes to the fauna.

Biases in the Data

In examining the data that has been collected, one can immediately recognize a strong sampling bias — apparently the pre-1980 work did not sample the Quinnipiac River fish fauna very thoroughly. The new lake and stream surveys nearly doubled the known localities for seventeen of the most common species and added ten species to the watershed survey. Some of these added species should have been found previously in the watershed (e.g., striped bass, blueback herring, rainbow trout, carp, and creek chub). These most likely were not collected because of their large size or commonness. A few of the added species, however, are apparently increasing their ranges in Connecticut (e.g., gizzard shad, rock bass, and fathead minnow) and do likely represent new occurrences within the Quinnipiac River watershed. The gizzard shad may have naturally expanded its range northward along the eastern seaboard, but the others have most likely been introduced to new waters by human activities, such as stocking of game fish or release of baitfish.

The Fish Fauna

A compilation of the results from all previous surveys shows that sixty species of fishes are documented at one time or another from the freshwaters and estuary of the Quinnipiac River watershed (Table 1). It should be noted, however, that not all of these species are still found in the watershed and some species have drastically changed in abundance over the years. In addition, another thirty-eight saltwater species found just beyond the mouth of the river in New Haven Harbor (Normandeau 1985) and known to enter estuaries (Whitworth 1996) could potentially be found in the lowest reaches of the Quinnipiac (Table 2). These were not documented because of limited sampling efforts within the estuary; with more efficient collecting in the estuary these species may be captured at one time or another.

Table 1 notes the distributions of fishes within the watershed. A distinction is made between estuarine and marine fishes because a few species (e.g., bay anchovy and tidewater silverside) are found primarily as estuarine residents (marked E in the table), whereas there are a number of fully marine species, which tolerate lower salinity water and also enter estuaries (marked M in the table). Likewise, a few species of freshwater fishes are tolerant of higher salinities (e.g., bluegill, white perch, white catfish, and black crappie) and are also found within the estuary (marked F/E in the table).

Four species that have not been found recently in the watershed include the bridled shiner, bluntnose minnow, pearl dace, and brook stickleback. The bridled shiner is a native species that appears to be on the decline in Connecticut (Whitworth 1996). In the 1930s this fish was found in North Farms Reservoir in Wallingford (Thorpe et al. 1942) and in the 1950s this species was abundant in many sections of the Muddy River according to collections in the Peabody Museum (YPM collections). It has not been recorded in the Quinnipiac since the 1960s, despite re-sampling in North Farms Reservoir and two sites in the Muddy River by the most recent lake and stream surveys. The other three species (bluntnose minnow, pearl dace, brook stickleback) were introduced and occurred in small isolated populations in the Quinnipiac watershed. Bluntnose minnows were found in the east branch of the Eight Mile River in 1952 (YPM collection). Pearl dace were found in a pond at Camp Clearview, just off the Muddy River near Route 150 in 1952 (YPM collection). The brook sticklebacks were known only from a branch of Patton Brook (Whitworth et al. 1968). One can reasonably conclude, then, that either these sites have not been properly re-sampled or events may have eradicated these species from the watershed.

Diadromous Species

Diadromous fishes move between freshwaters and marine for the purposes of reproduction and, depending on the direction of movement, are called either anadromous or catadromous. Catadromous fishes move out to sea to spawn, whereas anadromous move into freshwaters to spawn. A number of diadromous fishes were found in the lower portion of the Quinnipiac River, downstream of the Wallace Dam in Wallingford.

Wallace Dam, a former milldam with a low head (approximately five feet), represents the first major obstacle to fishes migrating upstream. Water continuously spills over the dam and, during periods of high streamflow, certain species of fishes may be able to surmount the dam. The catadromous American eel is widely reported from areas above the Wallace Dam and apparently has overcome this obstacle. Gizzard shad have been able to get above Wallace Dam, as evidenced by their presence in Community Lake (Jacobs and O'Donnell 1995). It is also possible that some river herrings (alewife, blueback, and American shad) are making their way in limited numbers above the Wallace Dam, but have not been found above the dam, because they must return to the sea after spawning. Other anadromous fishes, such as sea-run brown trout, striped bass, and Atlantic sturgeon, are not recorded from areas upstream of the dam.

All of the anadromous species, except for the sturgeon, were found in large numbers at the base of the dam. This indicates that these species could more easily conduct spawning runs further upstream if the dam were reduced as a barrier, by either constructing a fish ladder or completely removing the dam. Historical documents show that at least some of these species once ran in large numbers much further upstream. For example, American shad were reported in 1815 and earlier from an area just north of where the Ten Mile River empties into the Quinnipiac River in Southington (O'Hare 1981). One final anadromous species recorded from the watershed is the sea lamprey. Recently, this species was taken near the mouth of the Muddy River (Hagstrom et al. 1991). Sea lamprey may be more widespread within the watershed, but the larvae are difficult to sample without the use of an electroshocker.

An ammocoete larva of a lamprey was collected in the Eight Mile River in 1952 (YPM collection), but is poorly preserved and difficult to identify to species. The locality where this specimen was collected was, at that time, upstream of a significant number of dams, at least one of which is considered potentially impassible for sea lamprey (Stephen Gephard, Connecticut DEP, pers. comm.). That being the case, this specimen could be considered evidence that a population of brook lamprey may have existed in that location.

An anadromous species not yet recorded from the Quinnipiac River, but which should be found there, is the rainbow smelt. This species enters rivers in late winter/early spring for spawning. Very little sampling of the fish fauna in the Quinnipiac has been conducted at that time of year, so this species may have been overlooked. However, Normandeau Associates (1985) report the collection of eggs and/or larvae of rainbow smelt in the New Haven Harbor, just beyond the mouth of the Quinnipiac River.

Species at Risk

Atlantic sturgeon is listed as threatened throughout its range, including in Connecticut. In 1994, a fisherman caught a seven-foot adult individual at the base of the Wallace Dam in Wallingford (N. Hagstrom, Connecticut DEP, pers. comm.). It is highly unlikely that this species is spawning in the Quinnipiac River, instead, this large individual wandered upstream to the dam. This species has been caught occasionally in New Haven Harbor (Normandeau Associates 1979, 1985) and subadult fish are also known to infrequently ascend other rivers in Connecticut (Stephen Gephard, Connecticut DEP, pers. comm.). However, all former spawning populations in Connecticut are now considered extinct.

The bridled shiner appears to be significantly declining in both numbers and distribution across the state (Whitworth 1996). As reported above, this species has not been collected from the Quinnipiac drainage since the 1960s, despite the fact that it was at one time abundant in areas, such as sections of the Muddy River. This species should be considered for listing as a species of special concern in Connecticut.

Introduced Species

Twelve species (brown trout, rainbow trout, carp, goldfish, bluntnose minnow, fathead minnow, pearl dace, white catfish, rock bass, bluegill, largemouth bass, black crappie), most of which are food and game fishes, have been introduced into the watershed for recreational fishery purposes. Some of the stocking of introduced species goes back many years (William Hyatt, Connecticut DEP, pers. comm.). The U.S. Fish Commission and its later incarnation, the U.S. Bureau of Fisheries, in the 1870s to 1890s promoted the introduction of various trout and carp across much of the United States (Stroud 1975). Within the Quinnipiac River drainage, the state of Connecticut continues to stock rainbow, brown, and brook trout in the Muddy River, Ten Mile River, the "Gorge" area of the Quinnipiac River in Meriden, Wharton Pond, and Baldwin Pond. In addition, Thorpe et al. (1942) report that bluegills ("roach"), black crappie ("calico bass"), and largemouth bass were introduced to North Farms Reservoir prior to 1935. After a major anoxic event wiped out the fish fauna in that reservoir in 1935, the state quickly reintroduced both native and these non-native fishes to the lake. In many areas, these introduced species numerically dominate the fish faunas and likely have had a tremendously detrimental impact as competitors or predators on the native fishes (Whitworth 1996).

Carp were first introduced to the United States from France in 1832 by private citizens and brought over from Germany in 1872 by the U.S. Fish Commission (Lachner et al. 1970, Stroud 1975). Initially, they were considered a highly prized food fish and the Fish Commission encouraged their release across the U.S. However, by the turn of the century public sentiment had soured and the carp became viewed as a nuisance. Since then, carp have become even more widespread because of high fecundity and ability to tolerate a wide variety of environmental conditions. Carp are considered a pest species because they compete with other species for food, prey on other fish's eggs, and stir up the bottom sediments while rooting around for food, which diminishes the quality of habitat for other species (Tomelleri and Eberle 1990).

In addition to the more obvious introductions of non-native species of fishes, some species native to this region have been introduced to local areas where they didn't previously occur. For example, the creation of the Farmington Canal and its operation from 1828-1848 required the movement of water through the lock system. This not only transported water, but also fishes, between the Quinnipiac and Connecticut River watersheds. Whitworth (1996) attributes the movement of chain pickerel into the Quinnipiac River drainage basin and creek chub into the Farmington River drainage basin as a direct result of the operation of this canal. In many cases, these more subtle changes in the faunas occurred in the early 19th century in Connecticut and are more difficult to document (Whitworth 1996). Nevertheless, the effects were significant. For example, the introduction of the chain pickerel apparently led to a decline in the numbers of native brook trout in the Quinnipiac drainage (Whitworth 1996), presumably through predation on the young trout.

Historical Perspectives on Fish and Fisheries

What little we know about the fish fauna of coastal Connecticut prior to the arrival of English colonists comes mostly from archaeological excavations of shell heaps and middens. Examination of middens has given researchers some insight into the species of fishes caught by Native Americans on the coast of Connecticut. The list includes sturgeon, sandbar and sand tiger sharks, bluefish, striped bass, tomcod, tautog, cunner, black sea bass, sheepshead, northern barracuda, and skates (Waters 1965, Moore unpubl. data), and represents a small subset of the actual fauna. The absence of river herrings, for instance, is noticeable, especially considering the fact that herrings were not only reported as a food but were also used to fertilize agricultural fields (Rainey 1936). The typically acidic soils found along the coast may introduce a preservational bias by dissolving away the thin and fragile fish remains in some middens. Evidence also points to the fact that at some sites in southern New England fishing implements, such as net weights, harpoons, or fish weirs were produced representing some of the many techniques used to capture fishes (Banks 1990). Earliest colonial records from New Haven indicate that the Quinnipiac Indians used fish weirs, which work best at capturing herrings and other fishes that move through the estuary.

It did not take long for the English colonists to learn to utilize the local fishes. The original 1638 Treaty of the Plantation of New Haven includes provisions allowing the Quinnipiac Indians to continue fishing in the river so long as they did not take fish out of the English weirs or frighten fish away from such weirs (Townshend 1900). New Haven colony records from 1647 include a statute concerning not disturbing the fishing weirs set by either the local Indians or the colonists (Hoadly, 1857). The New Haven Town Records (Dexter 1917) continue to contain such notations, sometimes specifying areas in the Quinnipiac River. For instance, an entry from August 4, 1651 grants Ephrahim Pennington a small island lying in the East [Quinnipiac] River, provided he did not hinder the setting of fishing weirs there.

After a brief period of general warming in the 1650s, the climate in New England and elsewhere began cooling in the 1670s, producing extremely harsh winters by the 1680s and 1690s (Kupperman 1984). This period is known as the Little Ice Age. Water temperatures were probably lower as a result and likely impacted the reproduction and recruitment of fishes, such as river herring, in the Quinnipiac River, as it did to other fish populations elsewhere at that time (Pope 1997). A correlation of lower larval and juvenile survival with lower water temperatures has been shown, with shad (Leggett and Carscadden 1978). The climate did not warm up again until the early 1700s and it is then that an increased interest in fish and fisheries in the Quinnipiac River becomes evident again in the town records.

In the 1730s and 1740s, the town council began granting exclusive use of certain fishing areas in the Quinnipiac River and allowing citizens to literally stake their claims on the fishes found there. The earliest entry in the New Haven Town Records Sept 21, 1731, grants Seth Heaton, Ebenezer Frost and Daniel Barns fishing rights to an area in the Quinnipiac estuary called Andrews Point (Powers 1962).

An entry from the New Haven Town Records, dated January 12, 1740, grants to Joseph Bassett and Daniel Tuttle the possession of a fishing site immediately downstream of a bridge called Pine Bridge and allows them to set their stakes on each side of the river. Pine Bridge, later called Mansfield's Bridge, was just downstream of the present day Broadway Bridge in North Haven. This fishing area, known as the "Bridge Grounds" became a well-known place for large hauls of shad in the late 1700s and early 1800s. At that time, an annual catch of 4,000 shad was considered average (Thorpe 1892). However, Thorpe (1892) does describe the "Great Shad Haul at Quinny Dam" upstream of Pine Bridge, where an incredible 3,000 shad were taken in just two days!

Such fishing grounds were valuable because salted or smoked shad (which probably included the three species of river herrings) were early American staples. To fish in these exclusive areas required purchasing the "rights" from a "fishing company," which usually consisted of a cooperative group of individuals that shared the work and expenses stemming from fishing their area. Fishing rights could be quite expensive; Thorpe (1892) states that rights to fish the Bridge Grounds cost $200 in 1800. The majority of important fishing grounds were found in the estuary from Quinny Dam (just upstream of Pine Bridge) downstream to the harbor.

Shad were so important to the well being of the colony that accommodations for the fish were sometimes instituted, even to the detriment of other businesses. For example, Aaron Day in 1762 applied to the town fathers for permission to build a gristmill just upstream of Pine Bridge (Powers 1962). This planned mill included building a dam (later called Quinny Dam) across the river. The resulting accumulation of water would supply power to the mill. As a stipulation for approving the plan, the town fathers required the builders to leave a 15-foot gap in the middle as a "shadway." From April through June this fishway was left open to allow passage of the shad (Thorpe 1892). Of course during this time there was no accumulation of water behind the dam and therefore no power for the mill. The mill operation was essentially shut down for three months each year for the benefit of the fishes and those who depended on them.

Fishing pressure quickly took its toll, however. One Swedish traveler in 1753 noted that in New England there was already a generally perceived decline in the number of herring (Pearson 1972). The traveler attributed the decline to "immoderate catching" and the increasing use of dams for mills, which prevented spawning migrations. By the latter half of the 1700s, typical agricultural practices in New England were resulting in deforestation and increased fluctuations in water levels in rivers and streams (Cronan 1983). These various factors adversely altered the ecological conditions in rivers, including the Quinnipiac. In response to concerns that the fishes in the river were declining, the state felt it necessary to institute various controls on fishing effort. The 1796 Acts and Laws of the State of Connecticut included provisions preventing any fishing in the East [Quinnipiac] River downstream of Mansfield's [Pine] Bridge during flood tides (when the anadromous fishes enter the estuary) and also from sunset Wednesday to sunset Thursday each week. This law was enforced with stiff fines ($14) for anyone caught fishing when prohibited.

Even so, a high rate of fishing continued on into the next century. In 1856, William D. Hall set up a production plant, called the Quinnipiac Company, in a remote corner of the Mount Carmel area of Hamden. This company extracted oil from menhaden (presumably caught in the Quinnipiac River and New Haven Harbor) and sold the remaining parts as fertilizer. By the 1880s the company was selling at least 500 tons of fertilizer a year (Blake 1888), which presumably put a great deal of fishing pressure on the herrings found in the harbor and river. Hartley

(1943) attributes a single purchase of 9,000 herring at Oyster Point in New Haven to this fertilizer company.

Not all dams constructed on the Quinnipiac River or its tributaries were as "fish friendly" as the Quinny Dam described above. Increasing damming of the rivers for power took its toll on particular spawning runs and limited the overall reproductive space available to the anadromous fishes. In the latter part of the 17th century, a number of low dams were built to power grist or saw mills. The next century saw the legal establishment of "mill rights" which gave preference to mill owners and industry; the environmental disruption caused by milldams was outweighed by the economic benefits they conferred (Poirier and Gradie 1996). President Jefferson's 1807 trade embargo and the War of 1812 prevented the importation of many household goods and stimulated local manufacture of formerly imported goods. The 18th and 19th century saw increasing use of more substantial dams for mills and factories. Industrial areas proliferated along the Quinnipiac River in Southington, Meriden, and Wallingford, (Butler 1969). All these new dams prevented upstream migrations by anadromous fishes and created impoundments, which favored more lacustrine faunas.

The commercial fisheries for finfishes in the Quinnipiac estuary ceased before the beginning of the 20th century, but the watershed continued to provide recreational fisheries. Because more remote upstream sections were and are less impacted by human activities, they continue to provide habitat for native brook trout up to the present. The state continues to stock certain areas of the watershed, particularly The Gorge at the Cheshire/Meriden line, with three species of trout. Various lakes and impoundments furnished habitat for other game fishes, such as largemouth bass, sunfishes, and pickerels (Thorpe et al. 1942, Jacobs and O'Donnell 1995). More heavily urbanized sections of the river declined in habitat quality, a fact that is reflected in the resident fish fauna, where some stretches have only the most tolerant species present (Anonymous 1992).

Effects of Pollution

As both factories and town populations increased along the river, more industrial waste and raw sewage were dumped directly into the Quinnipiac. Water pollution became so severe that the state General Assembly enacted special legislation in 1886 to prevent the city of Meriden from discharging raw sewage into the river. This resulted in the construction of the state's first sewage treatment plant in 1891 (Anonymous 1992). Despite this action, pollution still increased on the river because many communities continued to dump, at best, primary-treated sewage into the river. There was also no regulation of industrial waste, so that by 1914 the State Board of Health declared the river polluted and fish life had practically disappeared from near the mouth of the river (Anonymous 1992). By the 1960s portions of the Quinnipiac River earned a "D" rating for water quality, meaning it was suitable only for transportation of sewage and industrial waste, and for navigation, power, or other industrial uses (Brown 1968). The pollution was so bad along almost the entire length of the river that anoxic conditions occurred in some areas. As a result, the state stopped stocking fish in Hanover Pond and the town of Plainville stopped stocking Hamlin Pond because the fish couldn't survive (Brown 1968). It was at this point that the state and, shortly afterward, the federal government instituted water quality regulations.

With state and federal regulations on water quality, the conditions in the Quinnipiac River began to improve considerably from the 1960s. Industrial discharges into the river substantially declined. In 1984, Southington put into place the first tertiary sewage-treatment plant, which removed excess nitrogen and phosphate nutrients from the effluent. By 1992, all the communities along the river had tertiary treatment plants. Current issues include non-point source pollution and enforcing compliance with the regulations. The next big challenge for the Quinnipiac River stems from the withdrawal of groundwater by various communities within the watershed. This drawdown of the water table results in a reduced flow in many of the tributaries, which not only limits the volume of living space for fishes in the river, but also allows for greater warming of the shallower waters in the watershed. Thus the environmental conditions the fishes and other organisms face are once again changing considerably.

Conclusion

It is apparent that the fish fauna of the Quinnipiac River watershed has undergone many changes over time and most recently, the changes are the result of human activities. What we find today is the end product of centuries of fishing pressures, pollution, physical alteration of the river, and introductions of exotic species. Fishing pressures these days are limited to subsistence or recreational fishing. The transition from water power to steam and then electricity for industry reduced the prevalence of dams, a number of which have since failed or been removed. Water quality legislation has decreased pollution and improved the overall environmental conditions. Each of these have helped restore better habitat for the fish fauna, in many areas (Quinnipiac River Watershed Association 1995). Given all the factors that have modified the river and those that continue to influence it, resurrection of the original fish fauna is very unlikely. Instead, we should attempt to balance the various utilities of the river and its fauna so as to achieve a "happy, healthy human population living in an aesthetically pleasing environment in which both are continually changing" (Whitworth 1996).

Acknowledgments

My thanks go to Bill Hyatt, Rick Jacobson, Stephen Gephard, Ernie Pizzuto and Neil Hagstrom, all from Connecticut Department of Environmental Protection, who provided insights and information regarding the Quinnipiac River and the fishes in state waters. Ray Pupedis from the Peabody Museum donated fishes recently collected from a number of sites on the Quinnipiac. Jim Gies of the Southington Water Department and Mike Osiecki of the Meriden Water Division furnished information regarding fishes in the reservoirs within their respective areas. Sigrun Gadwa graciously gave access to literature and information from the Quinnipiac River Watershed Association. George Berghorn kindly provided information on fishes found below Wallace Dam and helped create the map of the watershed. Finally, thanks go to Emly McDiarmid who encouraged the writing of this paper.

BIBLIOGRAPHY

Anonymous. 1992. Quinnipiac River. Connecticut Department of Environmental Protection, Bureau of Water Management Fact Sheet, Hartford, CT, 2p.

Banks, M. 1990. Aboriginal weirs in southern New England. Conn. Arch. Bull. 53:73-83.

Blake, W.P. 1888. *History of the Town of Hamden, Connecticut, with an Account of the Centennial Celebration, June 15th, 1886*. Price Lee and Co., New Haven, CT, 350p.

Brown, E.E. 1968. A survey of water pollution of the Quinnipiac River Basin. MS thesis, Southern Connecticut State College, New Haven, CT.

Butler, A.W. 1969. An investigation of the placement of dams and uses of the resulting water power on the Quinnipiac river from the source to the meeting of fresh and salt waters. MS thesis, Southern Connecticut State College, New Haven, CT.

Cronan, W. 1983. *Changes in the Land. Indians, Colonists, and the Ecology of New England*. Hill and Wang, New York.

Dexter, F.B. 1917. *Ancient Town Records, Volume I. New Haven Town Records 1649-1662*. New Haven Colony Historical Society, New Haven, CT, 547p.

Dunbar, L.E. 1984. Environmental monitoring of the Quinnipiac River Estuary. Connecticut Department of Environmental Protection, Water Compliance Unit, 20pp.

Hagstrom, N.T., Humphreys, M. and Hyatt, W.A. 1991. A survey of Connecticut streams and rivers- central coastal and western coastal drainages. Federal aid to Sport Fish Restoration F66R-3. Connecticut Department of Environmental Protection, Hartford, CT, 117p.

Hames, R.L. 1975. An evaluation of the fishery resources of the Thames River watershed, Connecticut. Storrs Agricultural Experiment Station, Bulletin no. 435.

Hard, W. 1947. *The Connecticut River*. Reinhart and Co., New York, NY, 310p.

Hartley, R.M. 1943. *The History of Hamden, Connecticut, 1786-1936*. Quinnipiack Press, New Haven, CT, 497pp.

Hoadly, C.J. 1857. *Records of the Colony and Plantation of New Haven from 1638 to 1649*. Case, Tiffany & Co., Hartford, CT, 547p.

Jacobs, R.P. and O'Donnell, E.B. 1995. A survey of selected Connecticut lakes. Federal aid to Sport Fish Restoration F-57-R-13. Connecticut Department of Environmental Protection, Hartford, CT, 46p.

Jokinen, E.H. 1994. Patton Brook, Southington, Connecticut macroinvertebrate study five year summary 1990-1994. Dunham Place Well Field Wetland Monitoring Report. Prepared for Southington Water Works, Southington, Connecticut.

Kupperman, K.O. 1984. Climate and mastery of the wilderness in 17th century New England. Pp.3-37, in: *Seventeenth Century New England*. D. Hall and D. Allen (eds.). Colonial Society of Massachusetts, Boston.

Lachner, E.A., Robins, C.R. and Courtenay, W.R., Jr. 1970. Exotic fishes and other aquatic organisms introduced into North America. Smithsonian Contributions to Zoology 59:1-29.

Leggett, W.C. and Carscadden, J.E. 1978. Latitudinal variation in reproductive characteristics of American shad (*Alosa sappidissima*): evidence for population specific life history strategies in fish. Journal of the Fisheries Research Board of Canada 35:1469-1478.

Linsley, J.H. 1844. Catalogue of the fishes of Connecticut, arranged according to their natural families. Amer. J. Sci. 47:55-80.

McDonald, M. 1887. The Connecticut and Housatonic Rivers and minor tributaries of Long Island Sound. pp.659-667, in G.B. Goode et al., *The Fisheries and Fishery Industries of the United States. Section V, Vol. 1, History and Methods of the Fisheries*. U.S. Government Printing Office, Washington, D.C.

Merriman, D. and Thorpe, L.M. 1976. The Connecticut River ecological study: the impact of a nuclear power plant. American Fisheries Society, Monograph 1, 252p.

Moore, J.A. and Boardman, R. 1991. List of type specimens in the fish collection at the Yale Peabody Museum, with a brief history of ichthyology at Yale University. Postilla (Peabody Museum, Yale University) 206:1-36.

Moss, D.D. 1960. *A History of the Connecticut River and its Fisheries*. Board of Fisheries and Game, Hartford, CT, 15p.

Narendra, B.L. 1988. Notes on some early college museums in America, with particular reference to Yale. Discovery (Peabody Museum, Yale University) 21(1):16-17.

Normandeau Associates, Inc. 1979. New Haven Harbor ecological studies, summary report 1970-1977. Prepared for the United Illuminating Company, New Haven, CT. Normandeau Associates, Inc., Bedford, NH., 735p.

Normandeau Associates, Inc. 1985. New Haven Harbor ecological studies, summary report supplement 2, 1981-1984. Prepared for the United Illuminating Company, New Haven, CT. Normandeau Associates, Inc., Bedford, NH., 199p.

O'Hare, E. 1981. *Quinnipiac River Corridor. Preservation - Recreation Action Plan*. Regional Planning Agency of South Central Connecticut, New Haven, CT, 45p.

Pearson, J.C. 1972. *The Fish and Fisheries of Colonial North America. A Documentary History of Fishing Resources of the United States and Canada. Part II. The New England States*. NOAA Report 72040302, National Marine Fisheries Service, Rockville, MD.

Pereira, J.J., Goldberg, R., Clark, P.E., Ziskowski, J.J., Greig, R.A., and Hughes, J.B. 1994. Utilization of the Quinnipiac River and New Haven Harbor as a nursery area by winter flounder. Quinnipiac River Fund Final Reports 1993, pp.38-69.

Poirier, D.A. and Gradie, R.R. 1996. As the wheel turns: changing perspectives on property rights and Connecticut hydropower. Bull. Arch. Soc. Conn. 59:27-37.

Pope, P. 1997. Early estimates: Assessment of catches in Newfoundland cod fishery, 1660-1690. Pp.7-40, in: Marine resources and human societies in the North Atlantic since 1500. D. Vickers (ed.). ISER Conference Paper No.5, Memorial University of Newfoundland, St. Johns, Newfoundland.

Powers, Z.J. 1962. *Ancient Town Records, Vol. III. New Haven Town Records 1684-1769*. New Haven Colony Historical Society, New Haven, CT, 884p.

Quinnipiac River Watershed Association. 1995. Canoe and natural resource guide to the Quinnipiac River. Quinnipiac River Watershed Association, New Haven, CT, 22p.

Rainey, F.G. 1936. A compilation of historical data contributing to the ethnography of Connecticut and southern New England Indians. Bull. Arch. Soc. Conn. 3:1-89.

Smith, C.P. 1946. *Housatonic, Puritan River*. Reinhart and Co., New York, NY. 532 p.

Stroud, R.H. 1975. Exotic fishes in United States waters. Sport Fishing Institute, Washington, D.C. SFI Bulletin 264:1-4.

Thorpe, L.M., Deevey, E.S., Bishop, J.S., Webster, D.A. and Hunter, G.W. 1942. A fishery survey of important Connecticut lakes. Connecticut State Board of Fisheries and Game, Hartford, CT, 339p.

Thorpe, S.B. 1892. *North Haven Annals. A History of the Town from its Settlement 1680 to its First Centennial 1886*. Price Lee and Adkins Co., New Haven, CT, 422p.

Tolderlund, D.S. 1975. Ecological study of the Thames River estuary (Conn.) in the vicinity of the U.S. Coast Guard Academy, Report RDCGA-575, 135p.

Tomelleri, J.R. and Eberle, M.E. 1990. *Fishes of the Central United States*. University Press of Kansas, Lawrence, KS, 226pp.

Townshend, C.H. 1900. *The Quinnipiack Indians and Their Reservation*. Tuttle, Morehouse, and Taylor, New Haven, CT.

Waters, J.H. 1965. Animal remains from some New England Woodland sites. Bull. Arch. Soc. Conn. 33:5-11.

Whitworth, W.R. 1996. "Freshwater fishes of Connecticut, second edition." State Geological and Natural History Survey of Connecticut Bulletin 114:1-243.

Whitworth, W.R., Berrien, P.L., and Keller, W.T. 1968. Freshwater fishes of Connecticut. State Geological and Natural History Survey of Connecticut Bulletin 101:1-134.

Additional data unpublished :

Berghorn, G. Unpublished collections of anadromous fishes at the base of the Wallace Dam, Wallingford, CT, 1996.

U.S. Fish & Wildlife Service. Unpublished fish collections from the Quinnipiac River estuary, North Haven, CT, 1989.

Table One

Species recorded from the Quinnipiac watershed. Species are noted as either native or introduced. Occurrence in the watershed is denoted by: A = *anadromous; C = *catadromous; E = estuarine; F = freshwater; M = marine species also entering the estuary.

Family Petromyzontidae			
Petromyzon marinus	sea lamprey	native	A
Family Rajidae			
Raja erinacea	little skate	native	M
Family Acipenseridae			
Acipenser oxyrhynchus	Atlantic sturgeon	native	M

* these terms are defined in the text.

Family Anguillidae			
Anguilla rostrata	American eel	native	C
Family Clupeidae			
Brevoortia tyrannus	menhaden	native	M
Alosa sapidissima	American shad	native	A
Alosa pseudoharengus	alewife	native	A
Alosa aestivalis	blueback herring	native	A
Dorosoma cepedianum	gizzard shad	native?	A/F
Family Engraulidae			
Anchoa mitchilli	bay anchovy	native	E
Family Salmonidae			
Salmo trutta	brown trout	introd.	F/A
Salvelinus fontinalis	brook trout	native	F
Onchorynchus mykiss	rainbow trout	introd.	F
Family Esocidae			
Esox americanus	redfin pickerel	native	F
Esox niger	chain pickerel	native	F
Family Cyprinidae			
Cyprinus carpio	carp	introd.	F/E
Carassius auratus	goldfish	introd.	F
Notemigonus crysoleucas	golden shiner	native	F
Luxilus cornutus	common shiner	native	F
Notropis bifrenatus	bridled shiner	native	F
Notropis hudsonius	spottail shiner	native	F
Pimephales notatus	bluntnose minnow	introd.	F
Pimephales promelas	fathead minnow	introd.	F
Rhinichthys atratulus	blacknose dace	native	F
Rhinichthys cataractae	longnose dace	native	F
Semotilus atromaculatus	creek chub	native	F
Semotilus corporalis	fallfish	native	F
Margariscus margarita	pearl dace	introd.	F
Family Catostomidae			
Catostomus commersoni	white sucker	native	F/E
Erimyzon oblongus	creek chubsucker	native	F
Family Ictaluridae			
Ictalurus catus	white catfish	introd.	F/E
Ictalurus nebulosus	brown bullhead	native	F/E
Family Gadidae			
Microgadus tomcod	Atlantic tomcod	native	M
Family Phycidae			
Urophycis chuss	red hake	native	M

Family Fundulidae			
Fundulus diaphanus	banded killifish	native	F
Fudulus heteroclitus	mummichog	native	M
Family Atherinidae			
Menidia menidia	Atlantic silverside	native	M
Menidia beryllina	tidewater silverside	native	E
Family Gasterosteidae			
Apeltes quadracus	fourspine stickleback	native	F/M
Culea inconstans	brook stickleback	native	F
Family Syngnathidae			
Syngnathus fuscus	northern pipefish	native	M
Family Moronidae			
Morone americana	white perch	native	F/E/A
Morone saxatilis	striped bass	native	M
Family Centrarchidae			
Ambloplites rupestris	rock bass	introd.	F
Lepomis auritus	redbreast sunfish	native	F/E
Lepomis gibbosus	pumpkinseed	native	F
Lepomis macrochirus	bluegill	introd.	F/E
Micropterus salmoides	largemouth bass	introd.	F
Pomoxis nigromaculatus	black crappie	introd.	F/E
Family Percidae			
Perca flavescens	yellow perch	native	F
Etheostoma olmstedi	tesselated darter	native	F
Family Pomatomidae			
Pomatomus saltatrix	bluefish	native	M
Family Sciaenidae			
Cynoscion regalis	weakfish	native	M
Family Labridae			
Tautoga onitis	tautog	native	M
Family Pholidae			
Pholis gunnellus	rock gunnel	native	M
Family Triglidae			
Prionotus evolans	striped searobin	native	M
Family Cottidae			
Cottus cognatus	slimy sculpin	native	F
Family Bothidae			
Paralichthys dentatus	summer flounder	native	M
Scophthalmus aquosus	windowpane	native	M
Family Pleuronectidae			
Pleuronectes americanus	winter flounder	native	M

TABLE TWO

Species of marine fishes found in New Haven Harbor (Normandeau 1985) and known to enter estuaries (Whitworth 1996) that may eventually be documented from the Quinnipiac watershed.

Family Carcharhinidae	
Mustelis canis	smooth dogfish
Family Squalidae	
Squalis acanthias	spiny dogfish
Family Clupeidae	
Alosa mediocris	hickory shad
Clupea harengus	Atlantic herring
Family Osmeridae	
Osmerus mordax	rainbow smelt
Family Synodontidae	
Synodus foetens	inshore lizardfish
Family Merlucciidae	
Merluccius bilinearis	silver hake
Family Gadidae	
Enchelyopus cimbrius	fourbeard rockling
Family Phycidae	
Urophycis regia	spotted hake
Urophycis tenuis	white hake
Family Belonidae	
Strongylura marina	Atlantic needlefish
Family Cyprinodontidae	
Cyprinodon variegatus	sheepshead minnow
Fundulus majalis	striped killifish
Family Gasterosteidae	
Gasterosteus aculeatus	three-spined stickleback
Pungitus pungitius	nine-spined stickleback
Family Serranidae	
Centropristis striatus	black sea bass
Family Carangidae	
Caranx hippos	crevelle jack
Selene vomer	lookdown
Family Sparidae	
Archosargus probatocephalus	sheepshead
Stenotomus chrysops	scup
Family Sciaenidae	
Leiostomus xanthurus	spot

Menticirrhus saxatilis	northern kingfish
Micropogonias undulatus	Atlantic croaker
Family Labridae	
Tautogolabrus adspersus	cunner
Family Mugilidae	
Mugil cephalus	striped mullet
Family Ammodytidae	
Ammodytes americanus	American sand lance
Family Gobiidae	
Gobiosoma ginsburgi	seaboard goby
Family Scombridae	
Scomber scombrus	Atlantic mackerel
Family Stromateidae	
Peprilus triacanthus	butterfish
Family Triglidae	
Prionotus carolinus	northern searobin
Family Cottidae	
Myoxocephalus aenaeus	grubby
Myoxocephalus octodecemspinosus	longhorn sculpin
Family Cyclopteridae	
Cyclopterus lumpus	lumpfish
Family Bothidae	
Citharichthys arctifrons	Gulf Stream flounder
Etropus microstomus	smallmouth flounder
Paralichthys oblongus	fourspot flounder
Family Soleidae	
Trinectes maculatus	hogchoker
Family Tetraodontidae	
Sphoeroides maculatus	northern pufferfish

Figure One

Map of the Quinnipiac River watershed. Localities mentioned in the text are noted on the figure.

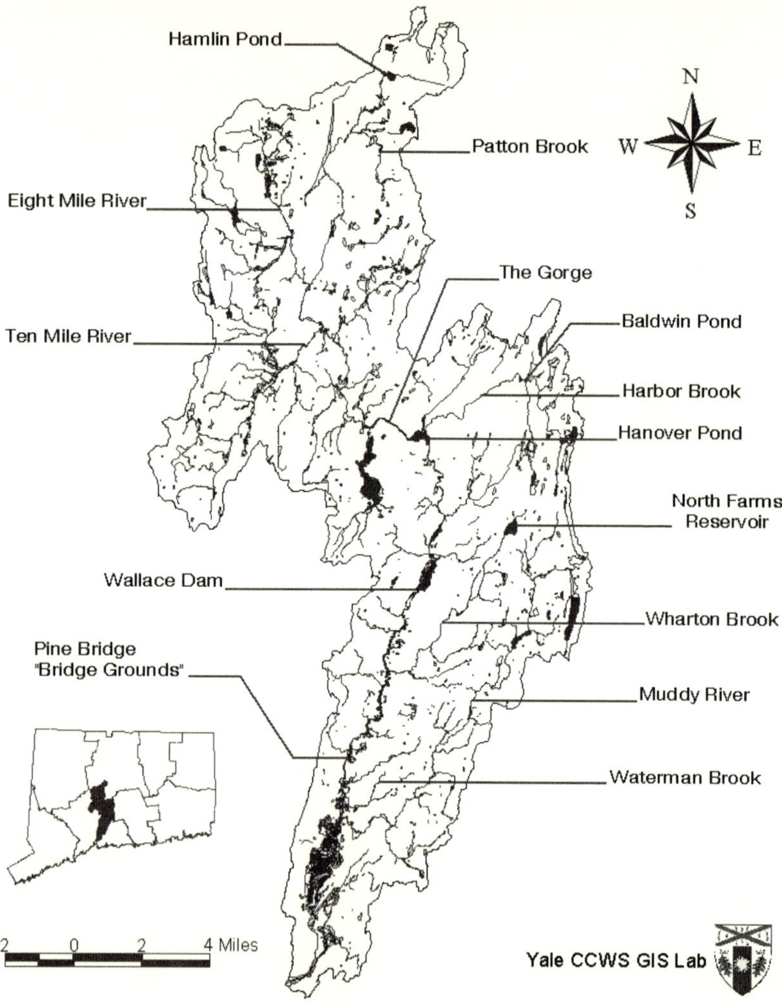